Joseph William Howe

The Breath and the Diseases Which Give It a Fetid Odor

Joseph William Howe

The Breath and the Diseases Which Give It a Fetid Odor

ISBN/EAN: 9783337811990

Printed in Europe, USA, Canada, Australia, Japan

Cover: Foto ©berggeist007 / pixelio.de

More available books at **www.hansebooks.com**

THE BREATH,

AND THE

DISEASES WHICH GIVE IT A FETID ODOR.

WITH DIRECTIONS FOR TREATMENT.

BY

JOSEPH W. HOWE, M. D.,

AUTHOR OF "EMERGENCIES;" CLINICAL PROFESSOR OF SURGERY IN THE MEDICAL
DEPARTMENT OF THE UNIVERSITY OF NEW YORK; VISITING SURGEON
TO CHARITY HOSPITAL; FELLOW OF THE NEW YORK
ACADEMY OF MEDICINE, ETC.

NEW YORK:
D. APPLETON & COMPANY,
549 & 551 BROADWAY.
1874.

PREFACE.

MARKED changes in the breath have heretofore received little consideration from the profession. Our medical text-books contain scarcely an allusion to their existence. I have endeavored, in the following pages, to give a succinct account of the diseased conditions in which a fetid breath is the most important feature. The principal facts detailed concerning the production of the offensive odors are the results obtained from personal investigations, conducted both in private practice and in our city hospitals.

J. W. H.

36 WEST TWENTY-FOURTH ST., }
September 1874. }

CONTENTS.

CHAPTER I.

PHYSIOLOGY OF REPAIR, DECAY, AND RESPIRATION.

CHAPTER II.

FETID ODORS FROM EMOTION, ETC.

CHAPTER III.

FETID ODORS FROM DYSPEPSIA.

CHAPTER IV.

FETID ODORS FROM BAD TEETH AND ULCERS OF THE MOUTH.

CHAPTER V.

FETID ODORS FROM CATARRHAL DISORDERS.

CHAPTER VI.

CATARRHAL ODORS (CONTINUED).

CHAPTER VII.

FETID ODORS FROM MINERAL POISONS.

THE BREATH.

CHAPTER I.

REPAIR AND DECAY—RESPIRATION.

General Considerations.—Sympathetic Connections.—Changes in the Food during the Process of Digestion.—Destruction and Repair of Tissue.—Compounds resulting from Physiological Decay and Putrefaction.—Composition of Air inhaled and exhaled.—Effects of Medicinal Substances on the Breath.—Causes of Fetid Breath.

AN offensive breath is a functional disorder liable to occur at all periods of life. Men are more subject to it than women. It is a prominent symptom of many morbid conditions. The affection derives its importance from the fact that it is a constant source of misery to all who, by force of circumstances, are compelled to associate with the unfortunate patient. In its worst forms it effectually destroys the communion of friends, and the pleasures of social intercourse. Even the harmony of the home-circle is invaded by a feeling of repugnance, which the best of us can scarcely control. Yet how few of the afflicted persons detect the cause of their isolation, or recognize the barrier which effectually prevents the approach of those near and dear to them! With the best intentions in the world, we rarely whisper a word of their disorder or suggest a source of relief. This false kindness—this demoralizing weakness—is universal.

In order to become acquainted with the sources of the fetor, to be enabled to prevent as well as to remove it, we must investigate some of the physiological processes manifested in the continual working of the animal organism. In a work like this such investigation must necessarily be limited. Enough can be learned, however, to give a proper understanding of the disorder in its various phases.

Each organ has characteristics peculiar to itself, and, though each process is distinct, there is no real isolation, for a multitude of nervous and vascular links bring all parts of the mechanism in close relationship. From the aggregation results one harmonious whole — a microcosm complete, and filled with exceeding beauty.

The nerves constitute a telegraphic system, through which the most delicate impression is transmitted from one part to another, with almost immeasurable rapidity, until it is registered in the great central office of the brain. The net-work of blood-vessels permeating the tissues throughout carries a constant stream of nutritive material to every part, neglecting none, and delivering its elements, according to the special needs of the organ it traverses. Thus the brain abstracts from the blood the peculiar food which sustains the nervous ganglia in their varied functions. The lungs remove from it carbonic acid, which has arisen from the molecular decay of tissue, and give to it a fresh supply of oxygen for sustaining the vital forces. The kidneys take away the excrementitious materials which constitute the urine. The liver receives from it the ingredients of the bile, as well as the nitrogenized substances, for its own sustenance. And so it travels, giving up one element here and another there, meeting constantly the diversified demands upon it, and creating a unity of action which makes up the life of the individual.

When morbid changes arise in the structure or in the functions of an organ, the whole body gives evidence of the pernicious effects. If the blood which circulates through the encephalic mass is deficient in nutritious elements, or poisoned by adventitious substances, the outflow or generation of thought is retarded, and the mental processes generally weakened. We wander from the subject —we forget. Our ideas are incoherent and pointless. There is a general feeling of depression, weariness, and anxiety. On the other hand, over-excitation of the brain overdrawing the reservoirs of nerve-force, weakens the power of digestion and impairs the quality of the nutritious elements taken into the system. Consequently the blood becomes impoverished, and in its turn increases and keeps up the disorder for an indefinite period. Again, when the heart is filled with the depraved blood, or when its nervous force is not sufficient to sustain its vitality, it becomes irregular in its action and beats with diminished force. The distant organs which depend upon it for a proper amount of vital fluid to keep them in working order are enfeebled, and fail to furnish healthy secretion. In this manner the morbid action is communicated to the whole human mechanism, until every fibre feels the change.

It is mainly, however, to a disarrangement of the functions of digestion and assimilation that we must look for the origin of "bad breath." The starches, sugars, fats, inorganic salts, and nitrogenized bodies, constitute the different elements necessary for alimentation, or for the support of life. The most important of these are the albuminoid bodies which are obtained from meat, bread, cheese, etc.

The albuminoid substances contain nitrogen, and hence are known as nitrogenized bodies. The great mass of muscular, osseous, nervous, and vascular tissues, in the body, is made of these substances in varied proportions. In the stomach, the fibrine and albumen of the meat, the gluten of the bread, the casein of the cheese, are mixed with the gastric juice and changed into a substance called *albumenose.* In this state they are absorbed by the blood-vessels, and carried by the blood, together with other portions of digested food, to the different tissues which they nourish and keep alive. With the process of supply there is also one of waste. Death of tissue is proceeding with a rapidity equal to the amount of repair. Life and death go hand-in-hand. Absolute death is essential to a renewal of life. The products of decay or "destructive assimilation" are carried off by the kidneys in the shape of urea, creatine, creatinine; by the lungs in the form of carbonic acid and water; by the bowels in the forms of excretine, and various gases, such as sul-

phuretted, carburetted, and phosphuretted
hydrogen. Some of the gases are manufactured in the intestinal canal by the decomposition of undigested food. Sulphuretted
hydrogen, which has an odor resembling rotten eggs, is found in greater abundance than
the other compounds of hydrogen. These
gases are secreted by the mucous membrane
of the intestines; sometimes they are thrown
off by the glandular apparatus in the skin.
This is often noticed when large doses of sulphur have been taken internally. In some
persons the quantity of sulphuretted hydrogen passing off by the skin is sufficient to
stain metallic articles carried in the pockets.
Outside of the body nitrogenized substances
undergo decay as they do inside. But in the
latter case the changes are so gradual, that
we usually fail to notice their extreme offensiveness. They pass through similar changes
whether decomposing in the intestinal canal
—the follicles of the mucous membrane—or
in cavities of the teeth. Coincident with the
process of putrefaction, fetid gases may be

secreted by the secreting apparatus of the intestines.

The air we take into the lungs with each inspiration is composed of seventy-nine parts of nitrogen, twenty-one of oxygen, a trace of carbonic acid and of ammonia. In certain localities adventitious gases are added, which render it exceedingly unhealthy. This is especially true of parts of New York, where street-cleaning has become one of the "lost arts," and, as a legitimate consequence, sewer and garbage effluvia are constant elements in the respired air.

The large amount of nitrogen in the air dilutes the oxygen, so as to divest it of all irritating qualities and make it respirable.

With each inspiration we take in only twenty cubic inches of air. This passes with each act as far as the first bifurcation of the bronchial tubes. As the capacity of the lungs is about two hundred and fifty cubic inches, it is therefore seen that but a small portion of the contents of the lungs is changed by each respiratory movement. There is, however, a constant interchange or diffusion of gases, by which the heavier oxygenated air is carried down to the air-cells, and the lighter carbonic acid carried up. The movement is assisted by the ciliated epithelium lining the walls. These cells "play" from below upward, thus creating a current at the sides, which assists the passage of the expired air through the bronchial tubes. The oxygen passes through the walls of the air-cells into the capillary ramifications of the pulmonary veins, and is carried off into the general circulation to the tissues in every part of the organism.

The products of decay or retrograde metamorphosis eliminated by the lungs, as before stated, are carbonic acid (composed of one atom of carbon and two of oxygen), water (formed of one equivalent of hydrogen and oxygen each, and a trace of animal matter). In a healthy state, when every organ is working naturally, there is no unpleasant odor from the expired air; but, as soon as the machinery gets out of order—as soon as extraneous materials are added—the breath is tainted. We may illustrate by taking the well-known effects of various volatile substances intruded into the system. Balsam-copaiba, in small doses, passes off by the kidneys; in large doses the lungs assist in its elimination, and its presence is then readily detected in the breath. Sandal-wood oil will affect the breath in thirty minutes from the time it is swallowed. Turpentine may be noticed in from thirty minutes to an hour. Sulphur will produce a very marked odor in the perspiration and expired air in two hours. The characteristic and more familiar odor of alcohol is easily recognized in the breath of an imbiber in ten minutes. These drugs pass off through the lungs when ordinary emunctories are overworked.

Prof. A. L. Loomis, of this city, has at the present time under observation an interesting case of *diabetes mellitus.* The patient has all the well-known symptoms of this disease, and, in addition, a breath which gives off a well-marked odor of sugar. In *diabetes*, there is a greater quantity of sugar manufactured than is required for the wants of the system. The surplus is thrown off, generally, by the kidneys. It is an exceedingly rare occurrence for the lungs to take part in getting rid of the sugar. In this case, however, there was probably an unusual amount made, or less of it assimilated, than usual; therefore, every organ capable of excretion was compelled to assist in throwing it off. In Bright's disease of the kidneys, *urea* is retained in the blood. This substance combines with the water in the circulating fluid to form carbonate of ammonia. In a short time the skin and the lungs endeavor to get rid of the poison, and then we can discern, in the perspiration and in the expired air, the characteristic odor of ammonia.

In the same manner, gaseous results of decaying nitrogenized tissue may be carried off when the other gate-ways are closed. A little sulphur in excess, combined with the hydrogen of the watery vapor going off during expiration, forms sulphuretted hydrogen, and causes at once an offensive breath. A simple rearrangement of the atoms of carbon and hydrogen will give carburetted hydrogen (CH_2), which also communicates to the breath a peculiar and disagreeable odor.

These facts being understood, it may be stated, as a general proposition, that any morbid condition of the system which prevents the elimination of metamorphosed nitrogenized tissues through the mucous membrane of the intestines, or retards the passage of the decomposing detritus through the bowels, will produce a fetid breath. The same result probably follows structural changes in the kidneys. Nature, to get rid of the poisonous accumulations, to maintain an equilibrium, must throw them off elsewhere, either in their offensive form or in modified, non-offensive combinations. Or, where the waste of tissue exceeds the repair, as in chronic, debilitating diseases and

low fevers, the eliminating glands are unable
to do the work imposed upon them, and vica-
rious elimination necessarily follows. Thus
the gaseous products of the stomach, the ex-
halations from the skin, become more or less
poisoned and fetid from admixture with foul-
smelling gases. The bad odor in the respired
air is more noticeable than from any other
part, but a close examination of the patient
will show that the skin likewise gives off
a very disagreeable fetor.

The various diseased conditions which
prevent the intestinal glands from eliminating the products of destructive metamorphoses are mental emotions, constipation, indigestion, congenital deficiency in the eliminating
glandular system, general debility, and low
forms of fevers. The local causes are decayed
teeth, caries of the nasal or maxillary bones,
ulceration of the lining membrane of the nose,
mouth, pharynx, layrnx, trachea, or bronchial tubes, or "putrid bronchitis." Chronic
poisoning from lead, arsenic, or mercury, may
also be enumerated as a common cause of

halitosis.[1]   In the subsequent chapters these
various diseased conditions will be considered, together with the appropriate treatment necessary for each.

[1] The term "halitosis" signifies diseased breath.   It is de-
rived from the Latin *halitus* (breath), and the Greek *nosos*
(disease).—*Harvard.*

CHAPTER II.

General Effects of Excessive Emotion.—Cases in which it destroyed
Life, and produced Serious Disorders.—Emotion as a Cause of
Bad Breath.—Class of Persons most subject to its Influence.—
Prevention.—Constipation, its Effects on the Respired Air and
Secretions.—Special Deodorizers.

THE influence of mental emotion on the
animal economy has never received the consideration which its importance demands.
Accustomed to regard the mind as some-
thing apart from the rest of the human
mechanism, we are apt to overlook it when
investigating questions of animal pathology;
and we constantly underrate its power over
the processes of growth and decay going on
within us. It is the tendency of the times
to accept only what we can see, feel, and
weigh, and gauge by our own narrow, in-
finitesimal conceptions the expansion of the

infinite. We separate too widely mind from
matter, and consequently come short of a correct appreciation of morbid physical changes.
This is not the place, however, for the illustration of this subject; nevertheless, it will
be necessary for illustration to give a few
instances of the effects of mental emotions
in totally changing healthy conditions, and
replacing them by serious and even fatal
disorders.

A criminal, who had been condemned to
death for murder, was given into the hands
of a celebrated French surgeon. He was
told that his judges had decided that he
should be bled to death on a certain hour
the following day. When the appointed
time arrived he was blindfolded and placed
in a bed. The surgeon then made a small
incision through the skin, which did not
involve any of the blood-vessels. One of
the persons interested in the experiment
placed his fingers on the pulse, another held
a vessel of lukewarm water above the wound-
ed arm and allowed the liquid to trickle over

its surface and drop on the floor. Meanwhile the surgeon, in his conversation with the assistants, alluded to the gradual weakening of the pulse, the fluttering of the heart, and the paleness of the countenance, until the criminal, fully convinced that his end was near, fainted and died, without having lost a single drop of vital fluid.

A similar incident occurred in Moscow a few years ago. A criminal, who had been condemned to suffer the death-penalty, was told that he was to sleep in a bed from which the dead body of a cholera patient had just been removed. He was then conducted to a well-ventilated room and placed in a bed perfectly clean, which had never been used. Toward morning he was taken with all the symptoms of cholera, and died in a few hours.

Instances of less violent results from mental emotion are not uncommon. Terror has changed the color of the hair from black to white. Fear may excite the sweat-glands to such an extent as to bathe the skin in a pro-

fuse perspiration. The same emotion may
so excite the action of the heart as to make
its impulse perceptible through the clothing.
A single thought will often take away the
appetite, or the remembrance of some favor-
ite article of diet will create one, and also
increase the secretion of saliva and buccal
mucus. Excitement has been known to
cause convulsions, dizziness, intense pain in
the head, and dimness of vision, which dis-
appeared on the return of calmer feelings.

Mental excitement may also alter the secretions of the alimentary canal and affect
the functions of the glandular system, so as
to prevent the proper elimination of the
disorganized, useless, nitrogenized materials
through the ordinary channels; and then,
as a natural consequence of the change, we
must expect an alteration in the composition
of the air passing off from the lungs, and a
contamination by gases, which give it a fetid
character. In such cases, the effect is brought
about simultaneously with the excitement;
in others, again, it develops subsequently.

At one moment the breath will be sweet, at the next fetid and unbearable. The odor is not so penetrating or disagreeable as that arising from decomposing food in the cavities of decayed teeth or from dead teeth. The patients are usually conscious of its existence, which is not the case in many other varieties of the affection. They are subject to violent fits of temper, easily excited, and as easily depressed. The appetite is variable. Sleeplessness is a common accompaniment. The following cases from my notebook afford good examples of the class under consideration :

CASE I.—Mrs. G., aged forty; occupation, singer. She is slight, but firmly built, and has an extremely sensitive nervous organization ; is able to endure fatigue without discomfort ; never had any hysterical manifestations, although extremely liable to lose control of herself under provocation. She has been subject at intervals to a bad breath, and is always conscious of its presence. It comes on suddenly, and is accompanied by a

2

peculiar taste in the mouth. It follows or accompanies fits of depression or anger, but never occurs from pleasurable excitement. Preceding the menstrual epoch it is aggravated; it disappears with the flow. Once in two or three months an intense headache comes on, with vomiting of bilious matter. This sometimes occurs without affecting the breath. She calls it a " feverish breath." Her appetite is good, and she is able to digest her food, except when suffering from depression. Constipation is also present at such times, but it *follows* and does not *precede* the bad breath, showing that the torpidity of the bowels is likewise an effect of the mental condition, and not a cause. She thinks the affection has lasted about five years.

CASE II.—Mrs. R., aged thirty-two. Is a lady above the medium height; inclining to *embonpoint.* She enjoys excellent health, has a good appetite, and digests her food readily. She is not subject to any special manifestations of excitement; fits of depression, however, are not uncommon at the menstrual epoch. The peculiarity in her case consists in this, that, whenever she is preparing to receive company, her breath becomes perceptibly "feverish." The odor continues until the excitement consequent upon the entertainment has subsided. Under similar circumstances the same thing has occurred for four years. During the free interval her breath is pure and sweet. An examination of the mouth and throat showed (as in the previous case) that there was no local cause for the affection.

Case III. occurred on the 23d of May last. On the morning of that day Mrs. R., a lady of forty-five, was bitten by a pet dog in the arm. She was very much alarmed at the occurrence, and fainted. I saw her an hour afterward, and, while examining the wound, discovered that her breath was very fetid. Having seen her frequently before without noticing any thing abnormal in the breath, I questioned her concerning it. She stated that she had been in perfect health up to the time of the accident, and that her breath had not been affected before it.

The suddenness with which the offensive breath develops in these cases may be explained by the theory spoken of above, or it may be that unusual nervous excitement induces a greater destruction of tissue than normally obtains, and, before the intestinal glands can accommodate themselves to the increased labor, other organs must assist in getting rid of the overplus of noxious material.

Bad breath, arising from mental emotion, is less amenable to treatment than any other variety, because it often depends on the will of the patient whether the cause shall be removed or not. There is no local fault which can be grasped and removed. The treatment, consequently, must be directed to the general system. All sources of mental irritation should be removed. Cold sponge-baths daily are of great service in giving tone to the nervous system. In certain cases shower-baths are good, but delicate persons are over-stimulated by them. Plenty of exercise in the open air is also a requisite. As much vegetable food as is consistent with a healthy state should be eaten at each meal, and the animal food diminished. When mental excitement is known to occur at certain stated intervals, the following preparation will be found of great service. It acts as a stimulant to the nerves, and enables the patient to pass the difficult date with ease :

℞. Tinct. lavend. comp.,	f ℥ ij.
Tinct. valerian,	f ℥ ss.
Mist. camph.,	f ℨ iij.
Aquæ carui,	f ℥ j.

Dose, fifteen drops on sugar every hour until the depression is relieved.

In some cases five or ten grains of musk, repeated once or twice during twenty-four hours, will be found efficacious.

Or either of the following may be employed :

℞. Tinct. valerianæ ammon.,	f ℨ ss.
Tinct. castor comp.,	f ℨ j.
Ætheris,	gt. xv.
Aquæ ancthi.,	f ℥ iss.

Half of this preparation may be taken two or three times each day.

℞.	Tinct. assafœtidæ,	f ℥ j.
	Tinct. hyoscyami,	f ℨ ij.
	Tinct. cinnam.,	f ℥ iss.
	Aquæ menth. pip.,	f ℨ ij.
M.		

Dose, one teaspoonful in water every three hours.

Should the breath be affected in spite of the preventive remedies, a wafer or pill, composed of the following ingredients, may be allowed to dissolve in the mouth. It effectually destroys the bad odor. Before using it, the teeth should be thoroughly cleansed and the gums sponged with a solution of myrrh and water:

℞.	Pulv. cinnam.,	
	Pulv. pimentæ,	
	Pulv. cardam.,	āā ℨ ss.
	Sacchari alb.,	ℨ j.
	Mucil. gum-acacia,	q. s.
Make fifty pills.		

Fifteen drops of oil of nutmeg, mixed with one teaspoonful of olive-oil, is a good deodorizer. It should be rubbed on the gums and cheeks with the finger.

Sometimes it is necessary to act on the

bowels with a mild cathartic before the odor can be removed. When the secretions of the intestinal canal are increased, the other medicines are more certain in their curative action. In such cases, a combination of rhubarb, myrrh, aloes, and oil of peppermint, acts beneficially. This compound passes under the name of "compound rhubarb-pill." Two or three may be taken at bedtime.

CONSTIPATION

Is more frequently a cause of bad breath than the preceding. A diminished secretion of mucus from the lining membrane of the intestinal canal, or deficient peristaltic action of the muscular coat, are its most common causes. Persons of indolent habits, who are accustomed to lounge in-doors, and rarely exercise in the open air, are very subject to it. A watery condition of the blood (*anæmia*) may produce constipation by lessening the normal amount of the secretion, or deteriorating the elements which enter into their composition. Nearly all forms of indiges-

tion have constipation as one of the prominent accompaniments.

When constipation exists, the disentegrat-
ed materials, which are thrown off through
the mucous membrane, and the remains of
undigested food, accumulate in the canal.
The mass is passing rapidly through all the
various stages of decomposition, and, by its
pressure and irritation of the glands, it diminishes the normal excretory function, and
Nature, to maintain an equilibrium, throws
its surplus of excretory materials on the
glands of other mucous membranes. Then,
again, the gases, in solution, arising from the
decomposing accumulation, may be taken up
by the blood-vessels, and thus increase the
effluvia.

It will be noticed, in most of these cases,
that the skin is darker than is natural, or
has a yellowish, semi-jaundiced hue, and
that the odor of its secretion is perceptibly changed. The tongue is coated. The
mucous membrane of the lips may be dry
and parched, and, in certain places, its epithe-

lial covering peels off, leaving the membrane
denuded. The breath is fetid, but the patient is generally unaware of it. When noticed by the patient, he usually expresses the
fact by saying that the breath is "*feverish.*"
Other signs of general disturbance of the functions of the body may after a time appear;
but their consideration belongs to other de-
partments, and will not be touched upon here.

The treatment of this class of cases, when
properly carried out, is eminently successful.
The fetor can always be removed. In the
first place, however, the patient must be
made to understand that medication *alone*
will not produce any permanent improve-
ment. An entire change of habit is required.
Vigorous exercise out-of-doors should be constantly enjoined. Where this is not practica-
ble, sparring, dumb-bell, or club exercise, will
be of great benefit. Fruit, such as oranges,
peaches, pears, etc., should be eaten in the
morning, before or during breakfast. Fruit
eaten in this way is a valuable adjuvant
in keeping up proper action of the bowels.

Friction of the abdomen with a rough towel after a bath, or at other times, will be found useful in promoting the same object. The patient should select for himself the most nutritious and digestible articles of diet best suited to his individual palate. What will answer for one will often interfere with the digestion of another; therefore no special diet can be laid down. In the majority of cases it will be found necessary to act on the bowels with gentle cathartics. They should be repeated every second day until the mucous membrane is restored to its normal condition. Active purgatives are rarely needed.

The following tonic and laxative preparations are exceedingly useful :

R̶.   Pulv. aloes socot.,   grs. xij.
      Ext. nucis vomicæ,   grs. ij.
      Pulv. ferri sulph.,   Ɔj.

Make twenty pills. One pill to be taken two or three times each day before meals, until the evacuations are regular and natural in appearance.

If there be pain and flatulence in connection with the constipation, it will be necessary to combine carminatives with the cathartic medicine :

R. Pulv. aloes socot.,   grs. xij.
Pulv. rhei.,   Ɵij.
Ext. nucis vomicæ,   grs. iv.
Pulv. zingiberis,   ʒj.

Make thirty pills. One pill before or after each meal. Or the following may be tried :

R. Magnes. carb.,   ʒj.
Sp. lavend. comp.,   f ℥ ss.
Olei anisi,   Ɵj.
Aquæ menthæ pip.,   f ℥ vj.

One tablespoonful four or five times each day before eating.

The officinal mixture of rhubarb, soda, and peppermint, given in teaspoonful doses five times in twenty-four hours, is also good in flatulence. A piece of rhubarb chewed, and the juice swallowed, will often answer the same purpose.

As in the previous cases, the mouth and teeth should be cleansed thoroughly with the solution of myrrh and water (one teaspoonful of the tincture of myrrh to a wineglass of water), or the carbolic-acid solution (four grains to two ounces of water). The tincture of cinnamon, slightly diluted with water, can be applied to the interior of the mouth and gums with a brush. It will communicate its flavor to the breath for a considerable length of time. Pieces of charcoal-cake as large as a hazel-nut may be eaten half a dozen times during the day with benefit. These things, however, only diminish, but do not entirely dispel the offensive odor. As a permanent deodorizer, the wafer mentioned in a previous page may be employed in conjunction with the following:

R. Pulv. carui sem.,
   Pulv. coriander sem.,
   Pulv. cinnam.,   āā ℨ ss.
   Sacch. alb.,   ℨ j.
   Mucil. gum-acacia,   q. s.

Make fifty pills. Dissolve one in the mouth when necessary.

The root of *Cornus Florida*, or sweet-flag, when chewed, impregnates the breath strongly with its peculiar aromatic principle. A piece the size of a pea will answer. The odor obtained by masticating the leaves of the common partridge-berry (*Gaultheria pro-cumbens*) is exceedingly agreeable, and makes an excellent deodorizer ; or, one teaspoonful of tincture of gaultheria and one of tincture of myrrh, added to an ounce of water, will be found equally efficacious.

Cigarettes, in which are mixed small pieces of cascarilla or cinnamon bark, may be smoked by persons who are in the habit of using tobacco.

Where the constipation and bad odor are kept up by a general lax condition of the system, a course of tonic medicines *alone* will often suffice to bring about a cure. Vegetable tonics, as a rule, agree better with the stomach than mineral ones do. A mixture, composed of equal parts of tincture of wild-cherry and cinchona barks, is good. If iron is indicated, the following may be given :

℞.	Ferri pyrophosph.,	ℨ ij.
	Quinæ sulph.,	ℨ ss.
	Acidi sulph. dil.,	q. s.
	Glycerine,	℥ iv.

One teaspoonful in a wineglass of water four times each day.

CHAPTER III.

THE false mode of life characteristic of the nineteenth century, the hurry, excitement, and worry, are generators of innumerable ills. Their most common sequences are disorders of the digestive functions. Sedentary habits, overwork, and the disgusting practice of " bolting " food, assist them in sapping the vitality until Nature gives way under the strain. Men eat as if they wished to get the food into the stomach in its most indigestible form. Their object is to save time — to save time, perhaps, for the cigar or bar-room. This is a great mistake. No reasonable excuse can be made for such a vicious practice, for all who indulge in it know, to their cost, the penalty, which, in the shape of dyspepsia and its kindred evils, constantly threatens them.

The symptoms which characterize indigestion are, in some respects, similar to those connected with chronic constipation. In many instances, torpidity of the bowels is but a result of a poor indigestion; yet, as indigestion frequently occurs independently of constipation, and, as habitual constipation is often associated with excellent digestive powers, it is but fair to make a distinction between them. There are, however, in addition to the symptoms before enumerated, pain and weight in the stomach after eating, and eructations of gas and acid liquids into the mouth. The gas may disturb the stomach and intestines so as to cause considerable pain. It rarely has a bad odor. Diarrhœa may be present, though the bowels are usually constipated. There are often great nervousness and depression of spirits, amounting, in some cases, to melancholia.

The breath becomes fetid very soon after the commencement of the disorder. The offensive odor is worse in the morning and evening. Fatigue and nervous excitement invariably increase it. Immediately after eating, it is scarcely noticeable.

The odor may be due to the decomposition of undigested food in the bowels, or from failure of the excretory glands to throw off the products of disintegration, or it may arise because decay of tissue is increased by the lack of support, resulting from the diminished quantity of nutritious elements taken into the system. This deficiency of nutriment is a constant and necessary effect of indigestion; and, with the excess of disintegration and the general torpor in the glands of the intestines, there is a demand for increased work in other parts to assist in throwing off the detritus. The following cases show the course of the affection in its various phases:

CASE I.—Mary M. G., aged thirty-one; occupation, nurse. Was first taken sick May,

1870. At that time she began to lose her appetite and feel uncomfortable after eating. A heavy meal was always followed by a pain in the epigastrium and side. Solid food affected her more than any other kind. With the pain a burning sensation (heart - burn) was sometimes felt in the chest. Pressure over the stomach excited nausea and vomiting. Cold water taken at night caused a painful feeling of distention over the abdomen, and a profuse flow of saliva into the mouth. On rising suddenly from the reclining posture, her head would become dizzy and her sight dim. She attributed this to her weakness. In the morning her tongue was coated with a yellowish-white fur, and her breath was exceedingly offensive. She was not aware of this latter symptom until informed of it by a friend, but since then the odor has been very perceptible to her own olfactories.

On examination of this patient I found her face sallow and care-worn, and her breath very fetid. Her eyes were heavy, and the edges of the lids red. The skin was dry and rough, like that of one convalescing from a fever, and the odor arising from it was unpleasant. While under observation, she abstained from all solid food, and confined herself almost exclusively to a milk-diet. This resulted in complete relief from her unpleasant sensations. Afterward she went to the country, lived out-of-doors, and finally recovered, without the aid of any kind of medication.

CASE II.—J. C., aged twenty-seven; occupation, book-keeper. Has been employed in a hardware establishment. For two years past he has been in the habit of working from eight in the morning until ten or eleven at night. The only exercise he took was walking to the store in the morning. About two years ago he began to be troubled with flatulence. Every thing he ate seemed to turn sour on his stomach, and he was constantly belching up large quantities of gas, which had an unpleasant odor. The bowels were alternately constipated and loose. In

about three months from the beginning of
the complaint severe pain after eating was
added to his other symptoms. The pain was
burning in character, and almost unbearable
until relieved by getting rid of the gas.
He was in the habit of taking brandy after
each meal, because it seemed to assist in digesting the food. His breath, he said, was
feverish from the beginning; he noticed it
was much worse in the evening than in the
morning. There was a bad taste in the
mouth continually. He thought the fetor
was caused by the yellowish matter which
accumulated on the tongue and gums between meals. This coating was made up
principally of epithelial cells and fine gran-
ules. It gave off a slight fetid odor; but its
removal made little or no change in the character of the breath.

These cases are typical, and they constitute at least one-third of the practice of many
of our medical men.

The same general hygienic measures recommended for the cure of constipation, and

the bad breath accompanying it, are necessary in every form of indigestion. Fresh air,
nourishing food, change of habitation, rest
from work and worry, do more toward promoting a cure than the most scientific medication. Unfortunately, many of our patients,
from necessity as well as inclination, remain
under bad sanitary influences.

When there are acid eructations, showing
excessive acidity of the stomach, alkalies may
be administered. Carbonates of soda or potash, given in five or ten grain doses, and repeated at short intervals, will give temporary
relief. Lime-water is also good in certain
cases, especially when there is nausea.

To empty the bowels, and at the same
time stimulate the secretions, the following
powders may be given :

    ℞. Hydrarg. chlor. mite,   grs. viij.
        Pulv. rhei.,   x.
        Olei anisi,   gt. v.

Make two powders ; one to be taken at night, and
the other in the morning. If the patient is robust and
full-blooded, both powders may be administered together. In the majority of cases it is best to com-

mence with a cathartic before resorting to tonics or deodorizers.

A pure milk-diet, continued for several weeks, will often remove the indigestion, and with it the offensive breath. Pieces of cracker may be soaked in the milk, or an egg may be beaten up with it once or twice each day. While pursuing this course, all varieties of solid food, except bread and light crackers, should be avoided. The quantity of milk taken must depend on the inclinations of the patient. In most cases, it may be increased until four or five quarts are consumed daily. When the milk disagrees with the patient, a trial must be made of other digestible articles of diet until something·is found to suit the delicate stomach. The bowels may be regulated by the pills of iron, nux-vomica, and aloes, or by Kissingen-water, taken in the morning before breakfast.

For the relief of the distressing sensation of weight and fullness in the stomach after eating, caused by flatulence, the following is always useful:

℞.   Bismuth. subnitrat.,   ℥ ij.
      Pulv. zingiber.,   ℥ iij.
      Spts. lavend. comp.,   f ℥ jss.
      Aquæ,   f ℥ ij.
M.

One teaspoonful four or five times each day, or as often as the distressing symptoms prevail. The medicine must be taken in water. Should the liquid preparation not agree with the stomach, the bismuth and ginger may be tried in the form of powder. Strong peppermint-tea, made by adding half an ounce of the leaves to half a pint of water, is also serviceable.

If this plan of treatment is followed out faithfully for a few weeks, the offensive breath, with the other disagreeable symptoms, will be entirely removed. When the patient recovers, the physician should insist on a strict observance of all ordinary hygienic laws in order to prevent a recurrence of the disorder.

For the immediate relief of the fetid breath, any of the previously-mentioned medicaments (page 36) may be employed. The charcoal-cake alone, when eaten four or five times during the day, will often suffice to remove the odor. When the tongue is much coated, it must be scraped clean, and the whole mouth washed with a solution of myrrh-and-lavender (one teaspoonful of tinctture of myrrh to a wineglass of lavender-water), or with the carbolic-acid solution. Unless this is done, the coating will become decomposed and add to the general offensiveness of the breath.

CONGENITAL BAD BREATH.

There are a few unfortunate persons who are afflicted with a bad breath from their childhood. At first, the affection is attributed to indigestion. As the child grows, however, he exhibits the ordinary amount of flesh, strength, and capacity for digesting food that other children do; he sleeps well, and has a good appetite, and, in all other respects, appears to be in perfect health. There is not one appreciable cause of bad breath to be discovered in the working of any organ. A close examination of the patient will show that the skin gives off a fetor similar to that coming from the lungs,

but much less intense. In warm weather, or
when under the influence of alcoholic stimulants, the breath of the patient is almost unbearable. Like some other forms, it is sometimes worse at night than in the morning.

It is probable that, in these cases, there
is some defect in the eliminating apparatus
of the intestines, or a peculiar tendency to
the formation of one class of metamorphosed
decaying substances instead of another. If
this view is correct, we may have a portion
of the carbonic acid and vapor of water, carried off from the lungs, replaced by some of
the fetid gases mentioned in the first chapter. Similar changes may also occur in the
secretions of the skin and other organs.

The treatment of congenital bad breath
is more palliative than curative. We cannot
remove the fetid odor completely; we can,
however, hide or modify it. The patient
should take a tepid bath daily, and afterward sponge the surface of the body with
cold water. When the skin has been rubbed
dry, a dilute solution of Florida-water should

be applied with a wet towel. An infusion of partridge-berry leaves, or cologne-water, may be employed in the same manner. A few drops of spirits of camphor act as a perfect deodorizer in some cases.

Every local cause which might add to the fetor must be removed. The mouth and teeth ought to be kept scrupulously clean. Charcoal, again, comes in here as a useful disinfectant. Sweet - flag, partridge - berry leaves, cinnamon - bark, etc., may be used constantly. A small portion kept in the mouth will hide the offensiveness of the breath. If necessary, all the deodorizers before recommended may be tried in turn, and the most suitable for the case selected for permanent use.

3

CHAPTER IV.

FOUL BREATH FROM DECAYED TEETH, ETC.

Decayed Teeth as a Cause of Bad Breath.—Effects of Decomposing
Food in the Cavities of the Teeth.—Causes of Decay.—Develop-
ment of a Vegetable Parasite in the Mouth from Uncleanliness.
—Accumulation of Tartar and its Chemical Composition.—Bad
Breath from Inflammations and Ulcerations of the Mouth.

PREMATURE decay of the teeth is often
witnessed in persons suffering from scrofula.
The teeth of such persons become discolored
at an early age, and crumble with very little
pressure. It is symptomatic of the general
lack of vitality, which exists in scrofula.
Teeth decay, likewise, from uncleanliness.
Small particles of food accumulate in the
cavities between the teeth, and in a short
time decompose. The decomposed material,
as it becomes packed and denser, presses on
the enamel, destroys it, and then has free ac-
cess to the soft bone beneath. In connection

with this deposit, a minute vegetable parasite is developed in the mouth, called the
Septothric bucollis. When examined under
the microscope, it presents the appearance
of a granular mass covered with filaments.
On the addition of iodine, a violet-color is
produced. In nearly every case of decayed
teeth, these parasites may be found. When
lodged between the teeth, they sprout with
the same rapidity that other fungous growths
do. It forms masses, which become harder
and denser as the development proceeds, until it absorbs the enamel in its neighborhood.
If a portion of teeth under it be examined,
its surface will present a roughened appearance, and, in the course of time, a large cavity is formed, opening into the interior of
the tooth. Another result of uncleanliness
is a deposit of *tartar* on the inner surface of
the teeth, near their insertion in the gums.
It adheres to the enamel like mortar, and, if
allowed to remain, causes destruction of that
part of the tooth upon which it presses. *Tartar*, which may be recognized by its hardness

and yellow-gray color, consists of phosphate of lime, mucus, salivary matter, and a peculiar animal substance, soluble in chlorhydric acid. The density and compactness of tartar are due to the phosphate of lime, the material which gives to bone its hardness. When, in the deposit, the animal matter is in excess, small parasites will often be found.

The use of acids for a length of time, either in the form of medicines or of unripe fruit, is another common source of decay which is often overlooked.

Decaying bone in all parts of the body, when exposed to the air, exhales a fetor. The same destructive process in the teeth occasions less odor than decay in other bony tissues. When a fetid breath is associated with rotten teeth, it is usual to assume that they alone are responsible; this, however, is not the case. The decaying structures of the teeth are but accessories in producing the fetor.

During the process of mastication, many particles of food become lodged in the minute cavities of the teeth. The heat and moisture of the mouth excite decomposition in the mass. The animal matters it contains, as well as the alkaline salts, assist in continuing the destructive process. In a day or two it is thoroughly rotten, and emits the foulness characteristic of putridity under other circumstances. Thus each decaying tooth becomes the receptacle and storehouse for decaying animal matter in its most disgusting forms. The portions of decaying material near the carious cavity become condensed and fastened into the crevices of the tooth. Daily the rotten mass receives fresh layers, which form on the top, until it is very difficult to distinguish the animal matter from the bone. As a necessary consequence of this putrefactive change, the breath becomes impregnated with foulness, so that it is almost impossible to remain in close proximity to the patient. The odor is worse than most other varieties, but it is soon relieved by appropriate treatment.

The proper person to consult in these

cases is the dentist. He should carefully cleanse the teeth from all incrustations of tartar, and remove the decaying substances from their cavities. These cavities may then be filled with what is known as "*soft-filling*," or with gold. Amalgam fillings are not conducive to health.. They do not make the breath any sweeter, nor do they preserve the teeth from decay.

The particles of food should be extracted from between the teeth after every meal by means of a toothpick. The teeth may then be washed with a solution of soap-and-water, applied with a soft brush. Hard brushes injure the enamel. The soap-solution destroys the parasitic accumulations of the mouth, besides acting to a certain extent as a disinfectant and deodorizer. When from any cause recourse to a dentist is impracticable, animal charcoal can be pressed into the cavities, allowed to remain for half an hour, and then be removed by washing. A solution of carbolic acid (two grains to the ounce of water) is an excellent deodorizer. It can be applied thoroughly when the charcoal is removed, and repeated if necessary. As the odor of carbolic acid is not always agreeable, its application ma*y* be followed by the myrrh-and-cinnamon solution :

℞. Tinct. myrrhæ,
Tinct. cinnam.,　　　　　　āā f ʒ ij.
Aquæ menthæ viridis,　　　ʒ ij.
M.

This mixture may be applied with a tooth-brush. The tincture of myrrh may be used alone, diluted with sufficient water to prevent it from irritating the mucous membrane. Either of the following may be tried :

℞. Tinct. calamus,
Tinct. gaultheria,　　　　āā ʒ j.
Spiritus myristicæ,　　　　ʒ j.
Aquæ,　　　　　　　　　ʒ ij.
M.

℞. Spiritus lavandulæ comp.,　　ʒ ss.
Aquæ menthæ piperitæ,　　　ʒ ij.
M.

These preparations may be used in a more diluted state if the patient is young. In some cases a diluted solution of nitric acid (ten drops to two ounces of water), used inside the putrid cavities, will destroy the animal deposit. Hydrochloric acid (fifteen drops of the diluted preparation to half an ounce of water) may be applied to the incrustations of *tartar*. This is done by dipping a piece of match or splinter of wood, prepared for the purpose, in the solution, and then rubbing it over the tartar until it is dissolved. It is better, however, to let the dentist attend to this also.

When the teeth and mucous membrane of the mouth are kept clean by these means, the offensive odor of the breath will disappear.

In all cases it is well to examine carefully the condition of the patient's health. When there is loss of appetite or general debility, fresh air and nourishing diet must be prescribed. A course of tonic medicines may also be found necessary.

Necrosis of the Jaw comes more within the province of the surgeon. The fetid odor connected with it is of small consequence in comparison with the pain and constitutional disturbance attendant upon it. It may result from decayed teeth, syphilis, or mineral poisons. When it is caused by dead teeth, usually but a small portion of the bone is involved in the morbid process. If it follow the introduction of mineral poisons into the circulation, the disease is far more extensive, and may destroy the whole of the bone. The necrosis produced by phosphorus is located generally in the lower jaw.

The treatment consists in removing the dead bone. When that is accomplished, the fetor will disappear. Should circumstances delay the operation, the bad odor can be diminished temporarily by applying pure nitric or hydrochloric acid to the surface of dead bone. This cannot be accomplished, however, unless the bone is denuded of its soft tissue, and then other deodorizers, such as carbolic acid, creasote, etc., must be resorted to.

Bad breath arising from putrid inflammation of the mouth (*stomacace*) is a peculiar and comparatively rare affection. It usually commences with congestion of the mucous membrane of the buccal cavity, followed by increased secretion. In a short time the gums begin to swell, and bleed from very slight pressure. Around the margins of the ulcers there is a yellowish exudation, which is thick and viscid, and not easily removed. The odor from the commencement to the termination of the disease is extremely offensive. It can be recognized at some distance from the patient. If the affection is not controlled by proper remedies, the cheeks become involved in the ulceration. Mastication and deglutition become very difficult, and the life of the patient may even be endangered. The disease is said by some to be contagious. It is often met with in children. The treatment is simple and usually successful. A solution of chlorate of potash, used repeatedly, in the course of a day or two stops the ulceration and removes the fetor. For adults, the following solution is the best :

℞. Potassæ chloras.   ʒ j.
     Syrup. zingiber,   ℥ ss.
     Aquæ,   ℥ iij.

One teaspoonful in water every three hours.

The mouth may be washed with the same preparation. Half the quantity of the adult dose may be given to children.

Scorbutic Ulceration of the Mouth is preceded by loss of flesh and strength. When the disease is fully developed, the gums become bluish-red, swollen, and painful. The tongue is coated with a yellowish exudation, and there is some increase in the quantity of saliva. Small extravasations of blood appear under the mucous membrane, and ulcers make their appearance, which bleed on slight pressure. In all these cases the breath is fetid, but the fetor is not so penetrating as that arising from *stomacace*.

The disease requires local and constitutional treatment. The local treatment consists in the application of astringent washes to the mouth, such as solutions of alum, tannic or gallic acid, uva ursi, and hydrastics.

One teaspoonful of powdered alum may be added to four ounces of water, and the mouth washed five or six times each day. The car-bolic-acid solution also is good—it corrects the fetor. The constitutional treatment con-sists in the administration of lime-juice, lemon-juice, and vegetable acids. The diet should be made up mainly of fresh vegetables, such as cabbage, potatoes, carrots, beets, etc., and a moderate amount of fresh meat.

Syphilitic inflammation and ulceration of the mouth and fauces are always accompanied by a bad breath. The ulcers rarely occur without signs of secondary or tertiary syph-ilis being present. There may be eruptions on the skin or periosteal inflammations at the same time. The ulcers are usually cir-cular and superficial. They are covered by a thin whitish or semi-transparent coating, and do not bleed easily. In their vicinity small, hard nodules or raised patches of mucous condylomata may sometimes be seen, or their margins may be indurated. The odor is not alone due to the secretion from the surface of the ulcer, but also to some change in the secretion of the salivary glands. Perhaps the change is effected by the matter from the ulcer acting upon the salivary ingredients.

The first thing to be done in these cases is to cauterize the sores thoroughly with nitrate of silver. A strong solution of carbolic acid and glycerine, in the following proportions, may then be applied over the ulcer and the whole mucous surface with a camel's-hair brush.

℞.	Acidi carbolic.,	℥ ss.
	Glycerine,	℥ j.
	Aquæ,	℥ ij.
M.		

Twice each day will be sufficient to make the application. The *black* and yellow wash (page 88) is often employed. Iodoform is an excellent remedy. The ulcers may be dusted with the powder two or three times each day. It is generally combined with glycerine (one drachm of iodoform to half an

ounce of the solvent), and applied with a
brush. Equal parts of myrrh and cinnamon
(one teaspoonful each), added to a wineglass
of water, can be used to rinse the mouth
when the odor is bad.

The constitutional treatment for syphilis
must be commenced at the same time (*see*
page 90), and continued until the ulcers are
healed.

"*Spontaneous Stomatitis*" is said to occur
suddenly, without any appreciable cause. It
is ushered in with soreness in the tongue,
gums, and teeth, and an increase in the flow
of saliva. Two or three days afterward there
is considerable swelling of the mucous membrane. The whole interior of the mouth is
of a deep-red color, and very tender on pressure. The flow of saliva becomes very great.
It runs from the mouth incessantly. It
is fetid, and communicates its fetor to the
breath. In some cases ulceration of the margins of the gums takes place, but this is
rare.

This disease resembles in many respects

the stomatitis excited by mercury. The diag-nosis is made by inquiring into the habits of the patient, and ascertaining whether mer-cury has been used or not.

Treatment.—The mouth should be washed with warm water containing a few drops of laudanum to allay the soreness. A solution of chlorate of potash (ten grains to the ounce) may be subsequently used on the inflamed surface, and repeated every three or four hours; or the potash may be combined with glycerine and belladonna (*see* page 72). Chlorate of potash is also useful as an in-ternal remedy. Ten grains three times each day will be sufficient.

Follicular Stomatitis occurs at all ages. The inflammation commences in the follicles of the mucous membrane on the inside of the lips, near their lower portion, and the sides of the tongue. Small red spots, which are firm to the touch and painful, first ap-pear. Softening of the membrane, in the centre of the hard mass, then takes place, and a small excavated ulcer results. In some cases the disease commences by a vesicular eruption (herpes). The vesicles heal, and leave small superficial ulcers. Hot liquids and solid food taken into the mouth cause much pain. The salivary secretion is increased, and the breath has an exceedingly disagreeable odor. This odor is present from the commencement until the termination of the disease.

As follicular stomatitis often depends on general debility, tonics are always indicated in the treatment. Combinations of wild-cherry, cinnamon, and cascarilla barks, are very beneficial. They may be given in the form of an infusion, half a pint of each mixed. A wineglassful, taken before each meal, will be sufficient. Tincture of gentian and calumbo (equal parts), given in teaspoonful doses three or four times each day, may also be tried. If the bowels are constipated, mild cathartics can be used. Soothing demulcent applications are required at the commencement of the inflammation. Mucilage, or sweet-oil, containing a little laudanum, is useful. Glycerine, with borax, is a common remedy, which brings good results. When ulcers form, they may be touched with nitrate of silver; or sulphate of copper, or oxide of zinc, may be sprinkled over them. The preparation of iodoform and glycerine is sometimes needed.

Gingivitis is an inflammation confined to the gums. It is more common in children than adults. The gums swell, become painful, and finally ulcerate. The ulcerative process is confined to the junction of the gums with the teeth. The breath is intensely fetid from the commencement of the ulceration. If the disease is not stopped by appropriate treatment, the teeth may loosen and drop out.

A strong solution of nitric or hydrochloric acid (twenty drops to half an ounce of water) may be applied to the ulcers and their inflamed margins two or three times in twenty-four hours; or the ulcerated surface may be cauterized with the acid nitrate of mercury or silver. Alum, in solution or powder, is also used. Chlorate of potash, employed as a wash, and taken internally, is one of the best remedies. Tonics should be given in all cases.

CHAPTER V.

CATARRHAL ODORS.

Clergyman's Sore-Throat (Follicular Pharyngitis).—Peculiarities of the Inflammation.—Effects of Decomposing Mucus on the Breath.—Complications.—Treatment.—Cancerous Diseases of the Tongue and Pharynx.—Fetid Odors following Diphtheritic Diseases of the Throat, etc.

CLERGYMAN's sore-throat is a chronic inflammatory affection of the mucous membrane lining the pharynx. The disease is located principally in the follicular pouches of the membrane. It occurs in persons who are broken down from overwork and sedentary habits, or who have been compelled to do a great deal of public speaking. In its advanced stages the disease is always accompanied by an offensive breath. It may arise from repeated colds, or it may extend from an inflammation of the lining of the nose or mouth. Sometimes the injudicious application of caustics to the membrane keeps up and increases the disorder. At first there is more or less soreness in swallowing. The voice becomes husky when singing or talking, and accordingly the act of speaking is painful. In the morning the patient feels as if a foreign body were sticking in his throat, and he endeavors to remove it by "hawking" or coughing. This sensation is caused by the thickened mucus which collects in the follicles, and on the surface of the membrane. After a time these follicles remain filled with the inspissated secretion. It becomes cheesy in consistency, decomposes, and gives off a very penetrating fetid odor that is communicated to the breath, and which remains as long as the disease exists. It takes from six months to a year for the disease to make a marked change in the sweetness of the breath; occasionally the breath is bad from the beginning.

On examination of the membrane, it will be found thickened, congested, and dark red in appearance. The blood-vessels in its substance are dilated and varicose. The follicles are choked up with a whitish material which can be picked out or scraped from the surface. It is from this substance that the odor comes. In some parts the epithelial covering which gives the membrane its smooth appearance is destroyed, and a rough, granular surface is seen, resembling the conjunctiva in "*granular conjunctivitis.*" In the majority of cases the mucous lining of the posterior nares will be found involved in the morbid action.

The cases recorded below are full of interest as examples of the usual course of this disease:

CASE I.—Martin C., aged thirty; married. Patient has been employed in a large railway-station, in this city, for some years. His work consists in calling the stopping-places before the departure of each train. He has been in the habit of using tobacco to excess for some years. Two years ago he noticed that his voice was becoming weaker, and that, when leaving work in the evening, he was extremely hoarse. A soreness followed the hoarseness, which has continued with but little intermission ever since. At times there is a sensation in the throat as if a pin or piece of bone were sticking there. In the morning it takes several minutes' hawking and coughing before he can clear his throat. Sometimes it is hard to swallow. He takes cold very easily, and every fresh attack is accompanied by a profuse expectoration. There is also a free discharge from the nasal passages. His breath has been affected for over a year. He did not know it until he observed that persons turned their heads away from him while in conversation.

Upon throwing a powerful light into the pharynx, I found that the whole mucous lining was studded over with grayish-white points, and between them the membrane was of a dark-red color. In some places the epithelium had been entirely destroyed. The tonsils showed similar changes. The offensiveness of the patient's breath was perceptible at the distance of a yard.

Case II.—Mary A., aged twenty-two; is occupied as a school-teacher eight hours during the day. The room in which she teaches is very close and warm. Nine months ago she noticed a tickling sensation in the throat, which was soon followed by a hacking cough. The cough became worse toward the close of school-hours. The voice was clear in the morning, but at night it grew very hoarse. A short time before presenting herself for treatment, she expectorated large quantities of yellowish mucus, which contained small, firm masses of cheesy matter, that had a bad odor.

On examination, I found the pharyngeal mucous membrane of a deep-red color in some places, and covered with whitish patches in others. Her breath was extremely fetid, yet she was not aware of it. The mucus scraped from the surface of the membrane emitted a pungent fetor. The lungs were normal.

This variety of sore-throat is difficult to cure, because it is generally neglected until the mucous membrane has become infiltrated with the products of inflammation. But relief can always be afforded, and the offensiveness of the breath completely destroyed. As the fetor depends, in a great measure, on the decomposition of the contents of the follicles, these must be emptied and kept completely free from the abnormal exudation. This may be accomplished by means of gargles or by spraying the throat with various solutions. The instrument for vaporizing the liquids is easily obtained at any drug-store.

One of the best solutions for either gargle or spray is the following:

℞.	Potassæ chloras,	ℨ ss.
	Ext. bellad.,	f ℥ j.
	Glycerine,	℥ j.
	Aquæ,	℥ iij.
M.		

To be applied four or five times each day. The nose may be washed at the same time by means of a syringe or with the spray-producer.

Another excellent solution, where the secretion from the throat is very profuse, is made by adding one teaspoonful of the tinct-

ure of the chloride of iron and one of glycerine to barely a pint of water. The application should be made as often as desired;
or the following may be employed:

℞. Tinct. myrrhæ,   f ℥ ss.
    Tinct. hydrastus,   f ℥ j.
    Aquæ,   ℥ iv.
M.

The mixture to be well shaken, and applied as before.

When the fetor is considerable, the throat
may be cleaned first with a strong solution
of carbolic acid (ten grains to an ounce of
water). A weaker solution may afterward
be employed exclusively, with decided benefit. The deodorizers previously mentioned
are also useful. When the secretion is purulent, a solution of permanganate of potash,
thirty grains to five ounces of water, will be
found of great value. It is to be used in
the same manner as the other preparations.
If the mucous membrane be much thickened, and the epithelium destroyed, a strong
solution of nitrate of silver (four grains to
the ounce) should be applied to the surface
4

daily for a couple of weeks, and then be followed by the preparation of chlorate of potash and belladonna previously mentioned.

These local applications must be persisted in for a long time, in conjunction with a course of tonic treatment. Tonics, fresh air, and exercise, are absolutely necessary to give strength to the system. So long as there is debility attendant upon the catarrh, local remedies have little effect.

The use of tobacco should be stopped, as well as all other habits injurious to the health.

If the disease has lasted so long as to prevent a cure, the constant use of deodorizers will be necessary to keep the breath in a good condition.

Syphilitic ulceration of the pharynx must be treated in the same manner as syphilitic ulcers in the mouth (*see* page 61).

Chronic enlargement of the tonsils is another source of offensive breath. The disease is apt to occur in young persons with the scrofulous diathesis. It often arises independently of neighboring inflammation. The offensive odor arises from accumulations of inspissated mucus in the follicles of the gland, and its subsequent decomposition. The odor is not as offensive as that produced by follicular pharyngitis.

The treatment consists in keeping the tonsils perfectly clean with the gargles previously mentioned. If the tonsils are enlarged, they may be painted with tincture of iodine, or with a dry solution of nitrate of silver.

DIPHTHERITIC ODORS.

Diphtheria and diphtheritic sore-throat. have an offensive breath as one of their most prominent symptoms. Especially is this true of epidemic diphtheria. But it is the sequelæ of these diseases, rather than the diseases themselves, that are under present consideration. When true diphtheria disappears, we find that the patient has a strong predisposition to catch cold and develop a sore-throat. Slight exposure suffices to inflame the mucous lining of the fauces and pharynx, and produce an offensive breath. As soon as the secretion of mucus increases, as a result of the morbid action, there is present a well-marked fetor, not so pungent, however, as that occurring in the original disease. In the simple forms of diphtheritic sore-throat there is also a fetid breath. It is, however, more easily remedied than the odor of true diphtheria. The thickened mucous secretion found in these cases is very thick and viscid, and it adheres tenaciously to the membrane. In some parts it is collected in lumps, which stick like glue to the tissues beneath. It is this · peculiar secretion which occasions the fetor.

As general debility is a common sequence of diphtheritic disease, our efforts should be directed to building up the system by means of tonics, nourishing diet, etc. In addition, astringent washes should be applied daily, such as—

R. Aluminis,   3 j.
     Aquæ,   ℥ iij.
M.
   Gargle the throat three times each day.

The muriated tincture of iron with glycerine, in the proportion of one drachm of
iron to an ounce of glycerine, may be applied to the inflamed membrane, with a camel's-hair brush, three times in twenty-four
hours. A nitric or muriatic acid solution
(ten drops to an ounce of water) applied
with a spray-producer, or used as a gargle,
will often act beneficially in restoring the
normal secretions and diminishing the bad
odor. A combination of carbolic acid and
tincture of iodine, in the following proportions, is recommended by Dr. Beverly Robin-
son :

R. Acidi carbolic.,   grs. x.
   Tinct. iodinii,   f ℥ ij.
   Glycerine,   ℥ j.

These preparations may be applied with a brush or
in the form of spray.

CHAPTER VI.

CATARRHAL ODORS (CONTINUED).

Ozæna depending upon Syphilis, Scrofula, Necrosis, and Caries of
the Nasal Bones, and Herpes.—Character of Fetid Odors.— Bad
Breath from Ulceration of the Larynx and Trachea.—Putrid
Bronchitis.—Bronchectasis, etc.—Treatment.

THE nasal mucous membrane is subject to
a variety of catarrhal affections. Some are
simple in their nature, and give but little
inconvenience. Others are characterized by
a profuse muco-purulent discharge, and a disgusting odor. The term *ozæna* is applied to
the latter class of cases. The fetor arising
from this disease is more offensive than any
other. According to French authorities, the
odor in it is similar to that arising from a
crushed bug. On this account they designate
the patient by the term *punais*.

Ozæna depends on a variety of causes,
such as syphilis, scrofula, herpes, foreign

bodies in the nasal passages, and death of the
nasal bones. The worst form of the disease,
however, has no assignable cause.

In syphilitic ozæna both nostrils are generally inflamed. In scrofulous affections, only
one nostril is involved. The discharge at
first is thin, irritating, and not very profuse.
If allowed to remain on the edges of the nose
and lips, it causes excoriation. After a time
the discharge is mixed with blood and pus.
The odor of the breath is extremely offensive to the patient, as well as his friends. It
sometimes fills the apartment in which he
rests.

If the interior of the nose be examined
with a speculum, some portions of the mucous lining will be found ulcerated, and the
bottom of the ulcers covered with a grayish-
white exudation. The remainder of the
membrane is generally much tumefied, exco-
riated, and of a dark-red color. In the course
of a few months the bones become involved,
and the nose is consequently depressed, producing a peculiar deformity, which is pathog-

nomonic of the disease. Other signs of secondary or tertiary syphilis are usually found in different parts of the body, which, with the history of the patient, lead to a correct diagnosis.

The following case affords a good illustration of syphilitic ozæna:

Jane C., aged thirty-six; occupation, seamstress. Contracted syphilis from her husband in the summer of 1868. She passed through the secondary form without much trouble. Six months afterward she noticed that her nose was sore, and that small scales formed on the inner surface. There was also a discharge which was extremely offensive, and which affected her breath seriously. She went to one of the city dispensaries and had it washed daily, and also took some tonic medicine. The treatment relieved her temporarily. During the winter of 1872 the nose again commenced to smell and be painful. The discharge was more profuse, and the breath more offensive, than at any previous time. Small pieces of bone came away in the discharges. When I first saw her, the discharge was of a dark color, and the odor of her breath was absolutely unbearable. The poor creature was herself conscious of its painful character. Her nose was flattened near the upper portion. In passing my probe into the nostril, I found that the nasal bones, or what was left of them, were movable and dead. The mucous lining was destroyed in several places. Her general health was also very much affected by the disgusting odor, the discharge, and by the consciousness that she was an object of disgust to her best friends. As she had no means, and could not be cared for at her own home, I sent her to the hospital, where she remained four months, and was finally discharged cured.

SCROFULOUS OZÆNA

Usually affects but one nostril. It is apt to attack young females who are broken down in health. It is preceded by all the symptoms of a common cold, such as sneezing, coughing, and increased secretion from the eyes and nose. In a short time the discharge becomes offensive, and affects the breath. It is sanious and watery, and excoriates the integument. It will often cease for several days, and then return with increased fetor. The fetor is not as offensive as in the syphilitic form of ozæna. An examination of the mucous membrane shows that it is swollen and corrugated. In some places there are fissures and ulcers from which a sanious fluid exudes. Occasionally the disease is situated so far up the nostril that its effects on the membrane cannot be seen. We then judge of the condition by the character of the discharge and the amount of fetor. The glands of the neck are sometimes enlarged, and there may be a deposit of tubercles in the lungs.

IDIOPATHIC OZÆNA.

The worst form of ozæna is one which commences without any assignable cause. It usually attacks robust individuals who are accustomed to the use and abuse of alcoholic stimulants. Young children are also liable to it. When once fully developed, it may last for years, resisting all attempts at cure.

The patient at the beginning of the disease is conscious of a "stuffed" feeling in the nose, like that resulting from an ordinary cold. A slight soreness accompanies the fullness. In a few days, the secretion from the mucous surface is increased. It is at first clear and ropy, but afterward becomes opaque and purulent. During the night the discharge may pass backward into the pharynx and down into the stomach. This produces nausea and vomiting. The breath is affected from the inception of the disease to its termination. The odor is sometimes so offensive that the patient is made sick by it. His wretched condition, only too apparent to his senses, causes great mental depression, and, unless relief is applied by medical treatment, he may become completely prostrated in mind and body.

The affection may last several months before ulceration of the mucous lining takes place. Some portions of the membrane may be thickened and others atrophied. The latter condition is most commonly noticed in the later stages. The ulcers which form are superficial, and rarely eat deeply into the substance of the membrane.

HERPETIC OZÆNA

Is a rare form of the disease. It may co-exist with an eruption of herpes upon the cutaneous surface. The affection is characterized in its early stages by a watery discharge from the nostrils, which is sometimes tinged with blood. There is considerable itching at the end of the nose. After a time, small hæmorrhagic crusts appear in the discharge, and the breath becomes affected. The odor is not so disgusting as that occurring in syphilis or scrofula. Still it is the most troublesome feature of the disease.

OZÆNA FROM POLYPI AND FOREIGN BODIES.

Polypi situated in the upper part of the nasal cavities may excite inflammation of the mucous lining, and a purulent fetid discharge. The growth of the polypus is accompanied by a sensation of fullness in the nose, and difficulty in breathing with the mouth closed. An examination of the nostrils will determine the location of the tumor. A cure usually follows its removal.

Children of tender years frequently insert peas, beans, and foreign substances into the nasal cavities, which enlarge by the absorption of moisture, and, by an increase of pressure, cause great irritation. Peas and beans have been known to sprout in the nasal cavities after having remained there several days, giving rise to serious inflammation of the mucous membrane and spongy bones. The discharge takes place generally from the nostril in which the foreign body is located. With the commencement of the inflammatory process the breath becomes more or less fetid, and continues so until the foreign body is removed.

OZÆNA ARISING FROM DEAD BONE

Is sometimes the result of injuries which destroy the vitality of the bone. The necrosis may also be due to scrofula or syphilis. A diagnosis is easily made if small pieces of bone drop out of the nostrils, or by passing a probe up to the dead structure. If the bone is dead, a rough, grating sensation will be communicated to the fingers.

Cases, however, occur in which the soft parts covering the bone remain intact for a long time after necrosis has occurred. This prevents it from coming away, or from being made perceptible with a probe. In such patients the nose, or the region occupied by the nasal bone, will often be swollen and painful on pressure. The pain, too, is generally worse during the night.

As the worst feature of all varieties of ozæna is the horrible odor of the breath, our first object should be to remove it and diminish the unhealthy character of the secretions. Remedies for this purpose are applied indiscriminately, without reference to the causes of the disease. The constitutional treatment, however, is directed to a removal of the cause, and consequently must vary in each case. Whether a complete cure is possible or not, the alteratives and disinfectants rarely fail to remove the disgusting features of the disease.

An ordinary glass or rubber syringe, a nasal douche, or " spray-producer," may be employed in cleaning the nose, or applying medicinal agents.

The nasal cavities should, therefore, first undergo a thorough cleansing with warm water, in order that the disinfectant may be applied directly to the diseased surface. If scabs, inspissated mucus, or pus, be allowed to remain in any part, the medicine will fail to affect the membrane underneath.

When the cleansing process is completed, the nasal douche may be filled with a strong solution of carbolic acid (eight grains to the ounce of water), and a steady stream of it al-lowed to pass through the nose for fifteen or twenty minutes. If the disease is situated at the upper portion of the Schneiderian membrane, the liquid can be forced up by compressing for a moment the nostril opposite to the one in which the nozzle of the nasal douche is placed. This process should be repeated at least four times the first day, so as to clean away every particle of bad-smelling material, and make a positive change in the secretions of the mucous membrane. After the second day, the operation may be repeated twice in twenty-four hours. If there is ulceration of the membrane, nitrate of silver may be used to cauterize the parts before any disinfectant is employed. A solution of chlorinated soda is a very good substitute for the carbolic acid. It may be used in the proportion of one tablespoonful of the solution to half a pint of water. Either the black or yellow wash may answer in some cases, but their disinfecting power is inferior to either of those mentioned. The black-wash is made by adding one drachm of calomel to a pint of lime-water; the yellow-wash by adding half a drachm of corrosive sub-

limate to one pint of water. These mercurial washes are specially applicable to patients
suffering from syphilitic ozæna. Creasote-
ointment, or creasote in solution, is also advised. Powders consisting of borax and sugar, or of chlorate of potassa, may be blown
up the nostril. Warm water, obtained from
the White Sulphur Springs, is recommended
by some physicians as an alterative and seda-
tive to the ulcerated mucous membrane. As
sulphur-water is not as irritating as some of
the other medicaments are, it may be applied
to the nostrils five or six times each day.
The odor of sulphur, however, is not a good
substitute for the fetor of ozæna; it will be
well, therefore, to follow the wash by one
of the deodorizers mentioned at page 55.
The combination of carbolic acid and iodine
will sometimes destroy the fetor when the
carbolic acid alone fails. The strength of
the solution may be increased in proportion
to the intensity of the odor. Coffee has
lately been introduced to the notice of the
profession as a disinfectant. If the beans be

chewed, the breath becomes strongly impregnated with their characteristic odor. In ozæna, a cold infusion of coffee may be injected
into the nasal cavities. It will be necessary,
before using it, to strain carefully, in order
to remove the small particles of coffee exist-
ing in the liquid. Carbolate of zinc is highly
spoken of. It is employed in solution (five
grains to an ounce of water). Where there
is much excoriation of the integument at the
borders of the nostrils, cold cream, "glycer-
ine-cream," or sweet-oil, may be applied constantly.

When ozæna arises from syphilis, mercury and iodide of potassium must be employed, in conjunction with the local treatment. If it accompany secondary syphilis,
one grain of the protoiodide of mercury may
be administered three times each day until
the gums begin to feel sore. In the tertiary
form iodide of potassium, in from, five to ten
grain doses, may be given four times each
day. Tonics are also necessary.

When scrofula causes the ozæna, the patient must have plenty of exercise in the open air, nourishing diet, and tonics. Cod-liver oil alone, or combined with preparations of iodine or bromine, will be found of special benefit. If the stomach will not tolerate the oil, fresh cream may be substituted. Among the best tonics used in scrofulous and other kinds of ozæna, are the following:

R.   Tinct. ferri mur.,   f ℥ ss.
     Quinæ sulphatis,   f ʒ ss.
     Glycerine,   f ℥ iv.

Dose, one teaspoonful in a wineglass of water, four times each day, before meals.

R.   Tinct. sarsæ,   ʒ iij.
     Tinct. guaiac.,   ʒ ss.
     Tinct. cinch. comp.,   ℥ iv.

Dose, one teaspoonful five times each day before meals. Infusion of wild-cherry bark, and infusion of catechu, in tablespoonful-doses, are also of benefit.

The idiopathic form of ozæna requires, in many cases, the same tonic treatment and local applications.

Herpetic ozæna is treated at first by strong alkaline solutions, such as—

℞. Liq. potassæ,   f ℨ ij.
   Aquæ,   ℨ ij.

or—

℞. Liquor sodæ chlorinatæ,   f ℨ ij.
   Aquæ,   ℨ ij.

These preparations should be applied to
the membrane in the manner previously de-
scribed. When the nose is clear, and the
scabs cease to be formed, deodorizers may be
applied.

Ozæna from foreign bodies cannot be
cured until every irritating particle is removed. If the patient is seen at the beginning of the inflammation, snuff, or other
sternutatory, may be introduced into the
nostril opposite to the one in which the ob-
struction is lodged, in order to induce sneezing. This method will often dislodge the
foreign body, and force it out of the nostril;
or a stream of water may be thrown into the
nostril with the nasal douche, in order to
wash it out. When these simple measures
fail, a long curved polypus forceps may be
passed up carefully to the foreign body,

closed upon it, and drawn down. Subsequently the inflamed membrane may be treated as in the previous cases.

Ozæna from dead bone can only be cured by removing the irritating material. It may be reached with a forceps through the nostril, or an incision may be made under the upper lip, behind the root of the nose, and carried upward until the dead bone is reached. The originator of this method is Dr. Ronge, of Lausanne, Switzerland.

HALITOSIS FROM ULCERATION OF LARYNX, TRACHEA, BRONCHIAL TUBES.—BRONCHIECTASIS.

There are two varieties of laryngeal ulceration liable to affect the breath; these are the tubercular and syphilitic. The latter is more frequently accompanied by fetor than the former.

Syphilitic ulceration of the larynx is an accompaniment of either secondary or tertiary syphilis. It is usually found in connection with inflammation of the periosteal covering of bones, nocturnal rheumatism, gummy tumors in various parts, and other signs of tertiary syphilis. The breath becomes offensive when the ulcers are fully formed, rarely before. The fetor is worst in the morning.

In the tubercular form of the disease, there will be a history of long - continued cough, expectoration of blood, emaciation; and there will be signs of tubercular deposit at the apices of the lungs. The breath is more offensive in the evening if hectic fever is present. It is not so unbearable as that arising from syphilis.

The offensive breath from syphilitic ulceration may be controlled, during the healing of the ulcers, by the inhalation of various disinfectants. The following solution may be placed in a large-mouthed bottle, and the vapor inhaled for two or three minutes at a time. If sufficient vapor does not arise from the heat of the hands, the bottle may be held over a spirit-lamp while the inhalation is taking place :

R. Tinct. iodinii comp.,   f ℥ ss.
   Aquæ ammoniæ,   f ℨ ij.
   Spts. vini rectif.,   f ℥ j.
M.
Shake well before using.

Another method of inhalation successfully employed is to cover the patient's head with a cloth, place the dish with the solution under it, close to the mouth, and then insert into the liquid a hot piece of iron wire to vaporize it. Dr. John A. Ripley uses a paper funnel for the same purpose. The small end is placed in the patient's mouth, and the broad end held over a hot shovel, upon which have been placed sub. sulphur and mercury. The steam-atomizer is often of great service in these cases. A solution of carbolic acid, or of any other disinfectant, is placed in the chamber. The lighted lamp at the bottom in a few minutes generates steam, and an extremely fine vapor is thrown from the mouth-piece. The inhalations may be repeated five or six times each day, unless they produce too much irritation.

For the relief of the offensive breath in tubercular ulceration, any of the deodorizers advised in the second chapter will be found efficacious.

Putrid Bronchitis is a rare affection. We know very little concerning its origin. In connection with all the ordinary signs of bronchitis, the breath is exceedingly offensive, and the matter coughed up from the bronchial tubes has also a fetid odor.

This disease bears some relationship to idiopathic ozæna; the fetid secretion probably originates in a similar way in both diseases. The remedies for the odor are the same as those recommended in syphilitic ulceration of the larynx.

Bronchiectasis, or dilatation of the bronchial tubes, is sometimes a cause of fetid breath. The mucus accumulates in the cavity of the bronchus until it decomposes. When the cavity is full, the foul matter is expectorated in large quantities. Perhaps once in twenty - four hours the tubes are evacuated in this way.

To remove the fetor, use any of the disinfectants.

5

CHAPTER VII.

FETID ODORS FROM MINERAL POISONS.

Use and Abuse of Mercury.—Organs which eliminate the Drug.—
Effects on the Salivary Glands.—Quantity of Drug necessary to
produce Salivation.—Mercurial Fetor.—Remedies.—Bad Breath
from Arsenic, Lead, Antimony, Phosphorus, etc.—Treatment.

MANY years ago mercury was considered
a panacea for every ill. It was administered
in all forms of disease. Whether the patient
had fever, a chill, pain or numbness, wakeful-
ness or drowsiness, or whether he was full-
blooded or thin-blooded, short, tall, stout, or
emaciated, a plebeian or aristocrat, the mercury was given; and it was not considered to
have fully accomplished its work until the
" gums were touched." In those days, mercurial fetor and salivation were an ordinary occurrence. Even at the present time, cases of
poisoning by this drug are not rare.

Mercury is prescribed for all forms of syphilitic disease; for sluggishness of the liver, and constipation. It is generally carried out of the system through the kidneys. When the blood is overloaded with the poison, these organs fail to eliminate a sufficient quantity. The liver, salivary glands, the mucous membrane of the alimentary canal, and perhaps the lungs, then assist in its removal. The salivary glands throw off the largest amount. A large quantity of liquid is secreted by them in order to retain the mercury in a solution, and we have, as a result, the characteristic salivation.

Mercurial fetor may be produced by taking one grain of calomel three times each day for four or five days. It can be noticed when the soreness of the gums is scarcely appreciable. Blue mass, given in two-grain doses, with the same intervals, will affect the breath in seven days. If the kidneys are diseased, a much less time is necessary to produce fetor and salivation. In such cases

I have seen a cathartic dose of calomel cause salivation in twelve hours.

Mercurial stomatitis is apt to occur among persons who work in quicksilver-mines or looking-glass manufactories.

At the beginning of the disease there is noticed a disagreeble metallic taste in the mouth, which the patient likens to the taste of copper. There is a peculiar feeling of soreness experienced at the roots of the teeth when the jaws are closed. The gums are sore to the touch. A disagreeable fetor is communicated to the breath, both from the mucous membrane and the saliva. It is the most disagreeable feature of the disease. Shortly after, the saliva flows profusely. A grayish white line appears around the edge of the gums. The gums swell and may ulcerate. The teeth often become loose. If the disease is not removed by proper remedies, the tongue and cheeks become involved, and swell up so that eating is almost impossible. This condition adds an element of danger to the case.

Treatment.—If the patient is taking mercury as a medicine, it should be stopped. Or if he has developed the disease working in quicksilver, another employment must be chosen. The mouth should be washed thoroughly four or five times each day with a strong solution of chlorate or permanganate of potash—the chlorate is the most suitable. Glycerine and borax may also be used. A teaspoonful of powdered alum in a wineglass of water is a good application when the flow of saliva is very great. Half a teaspoonful of tincture of opium in an ounce of mucilage is an excellent anodyne if the soreness is great. Belladonna may be substituted for the opium if desired. Twenty grains of tannic or gallic acid, dissolved in an ounce of water, may also be used to diminish the secretions. If these various washes do not destroy the disagreeable odor of the breath, any of the deodorizing liquids or pills, previously mentioned, may be used.

The internal remedies for chronic mercurial poisoning are iodide of potassium and

chlorate of potash. The former is more frequently used. Either the iodide or chlorate
may be given in five or ten grain doses, four
times each day. The iodide of potassium
is supposed to join with the mercury in the
blood and tissues to form the soluble iodide
of mercury which is eliminated through the
salivary glands and other organs.

ARSENICAL SALIVATION

May result from a long, constant use of
Fowler's solution, or arsenious acid, or from
inhalations of microscopical particles of arsenic which arise from the green surface of
room-paper and artificial flowers. The system becomes saturated with the drug, and
Nature calls upon the salivary glands, the
skin, and other organs, to carry it off.

The fetor and salivation may be preceded
by a disordered condition of the alimentary
canal, such as nausea after eating, pain in
the epigastrium, and diarrhœa. The skin
presents the waxy pallor of Bright's disease.

Eruptions, such as eczema, also occur, and are probably due to the attempt of Nature to eliminate the poison through the skin. Early in the disease the breath has a disagreeable odor. It is said by some to resemble the garlicky odor obtained by burning arsenic. The fetor is increased by the indigestion, which is always an accompaniment. The salivation is not so profuse as in mercurial poisoning. Small ulcers may form at the edges of the gums, and on the mucous lining of the cheeks and throat. Where these make their appearance, the fetid odor is increased.

The administration of preparations of iron is advisable in all cases. As a direct antidote, a teaspoonful of the hydrated sesquioxide of iron may be given every three or four hours. Muriated tincture of iron is an excellent remedy in twenty-drop doses, largely diluted, every four hours; or the preparation may be given in combination with quinine and glycerine (*see* page 91). When mixed with glycerine, the iron is also useful as a local application to the inflamed mucous membrane. It diminishes, to a great extent, the fetid odor. Fresh air and nourishing diet are necessary auxiliaries in the treatment.

Half a teaspoonful of cajeput-oil added to half an ounce of sweet-oil is a good deodorizer to use in arsenical salivation. A small quantity may be rubbed over the gums and inside of the cheeks with the finger. The application may be made every two or three hours.

LEAD.

A fetid breath from lead-poisoning is a comparatively common occurrence. It manifests itself generally with the constipation which is one of the first effects of the poison on the system.

Chronic poisoning may be caused by using hair-dyes, drinking beer or water which flows through lead pipes, constant handling of the tin-foil covering chewing - tobacco, manufacturing or mixing white-lead. It is sometimes produced by wearing Brussels lace, the material of which owes its white color to carbonate of lead. The symptoms which follow the fetor, and constipation, are intense colicky pains in the abdomen; retraction of the abdomen, due to paralysis of the recti muscles; soreness of the gums, with a blue line around their edge ; increase in the flow of saliva, and " thumb drop " and " wrist drop " from paralysis of the extensor muscles. These symptoms vary in intensity with the amount of the poison taken into the system. Sometimes the only symptoms manifested are the disagreeable odor to the breath, blue line around the gums, and constipation. The existence of constipation adds, no doubt, very much to the fetor. The bad breath from lead-poisoning is more easily removed than that produced by any other mineral. It often disappears when the bowels have been completely emptied by appropriate cathartics. Two or three drops of croton-oil, mixed with mucilage or sweet-oil, is a common remedy where obstinate constipation exists. The dose may be repeated in an hour if the first fails. Elaterium may be given in quarter-grain doses, repeated at intervals of an hour until free evacuations are produced. Warm water, with an ounce or two of castor-oil, may be used in the form of an enema for the same purpose.

Iodide of potassium is considered by many the best eliminative. It joins with the lead in the system to form a soluble iodide of lead, which is carried out through the different evacuations. Sulphuric acid is often administered for the same purpose. The paralyzed limbs may be healed by friction, electricity, and cold water.

Any of the disinfectants mentioned on page 55 may be employed in these cases.

ANTIMONY

Produces a bad breath by increasing the waste and destructive metamorphoses of tissue, and by disordering the functions of digestion. When it has been taken in small doses for two or three weeks, the strength begins to fail. There are loss of appetite, nausea, pain in the abdomen, and looseness of the bowels. The pulse becomes very small and feeble. As soon as the kidneys and bowels fail to get rid of the drug, and the detritus of nitrogenized decay—the mucous membrane of the mouth and lungs—the skin and salivary glands both fail in their work, and the breath is made fetid. The foul odor is sometimes associated with a metallic taste in the mouth. The principal medicine employed as antidotes to antimony are the vegetable astringents, such as tannic and gallic acid, and strong infusions of green tea. One teaspoonful of tannic acid may be added to half a pint of water, and taken in repeated doses in the course of three or four hours. Green tea is probably the best remedy in chronic poisoning, because, in addition to its antidotal character, it acts as a stimulant. Where there is much nausea, mustard-plasters should be applied over the epigastrium, and iced champagne taken continually. Stimulants and tonics are always necessary to support the strength of the patient. Any of the preparations of iron and quinine may be employed with benefit.

PHOSPHORUS.

Chronic poisoning from this substance usually occurs among the employés in match-manufactories, from inhalation of phosphorous vapor. Like the preceding, it usually manifests itself first by dyspeptic symptoms, such as loss of appetite, weight, and heat in the epigastrium and prostate. The breath is affected early in the disease, not so much by the drug, as it is carried out of the system, as by the failure of nutrition and consequent increase in destructive metamorphoses. When necrosis of the jaw sets in, the fetor becomes unbearable.

Tonics, change of air, and exercise, are necessary to restore the general health. The constant use of disinfectants will be necessary until recovery takes place (*see* page 55).

THE END.

www.ingramcontent.com/pod-product-compliance
Lightning Source LLC
Chambersburg PA
CBHW031442280326
41927CB00038B/1505